A ROCKET TRIP

BY SANU SAMUEL
ILLUSTRATED BY DUANE HENNESSY

notionpress.com
INDIA • SINGAPORE • MALAYSIA

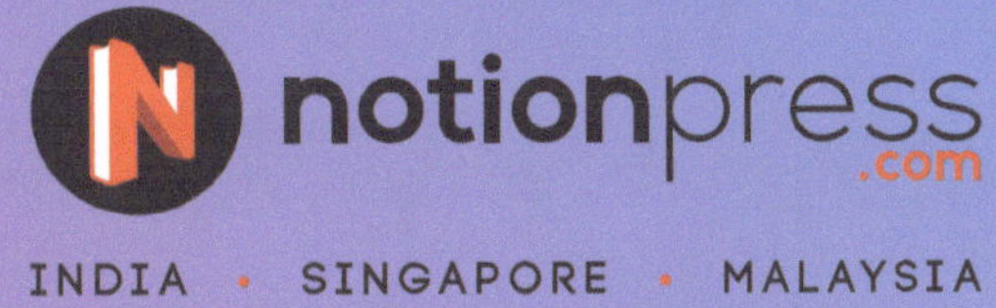

Jeremy is a young boy who loves to play outside.
One day he takes a trip on a rocket and experiences an
incredible journey.

An exciting story of a young boy's
adventure to space.

This book is dedicated to my dear family who
have always stood by me and supported me
during my good and difficult times. Special
thanks to my younger sister,
Sasa who has helped me make
this a reality.

This book belongs to

Name : ………………………………

Age :   ………………………………

There was a young boy named
Jeremy, who enjoyed playing
outdoors. He would spend time
in the garden
catching ants and snails.

One evening, as he was
cycling back home,
he looked up and he saw a
white trail in the sky.
He wondered if it
could be a shooting star,
a jet or a rocket
flying through?

He always wanted to take a
trip on a rocket and with that
wish in his mind, he
went to sleep.

The next morning to his big
surprise, he saw through his
window something he had
not expected.

A huge rocket stood in his garden!
He was so thrilled that he ran
outside screaming
with excitement.

He climbed up the
stairs excited to
reach the door of
the rocket.

As he entered the rocket, he could hear the computerised sounds that instructed him to wear a space suit, space helmet and get ready for the adventure of a lifetime!

A bit nervous, Jeremy put on his gear and buckled himself into the pilot's seat.

As soon as he sat, he could hear the door shut and a countdown begin.

EXIT
KEEP CLEAR
PLEASE PUT ON YOUR SPACESUIT AND SPACE HELMET AND GET READY FOR THE ADVENTURE OF A LIFETIME!

5 ... 4 ... 3 ... 2 ...1 ...
blast off and
BOOM the rocket shot
up with mighty speed to
the skies leaving a
greyish-white
trail behind.

BOOM!

Jeremy jerked a bit, while
the seat belts held him
tight. He was glad he could
breathe through
his space gear.

Within a few minutes Jeremy could see through the window of the rocket, the breath-taking view of the Earth.

WEATHER SATELLITE
FLIGHT KLF512
NAME: EARTH
LIFE SUSTAINABLE: YES
AGE: 4.6 BILLION YEARS
ORDER FROM THE SUN: 3RD PLANET
HIGHEST POINT: MOUNT CHIMBORAZO,
ECUADOR
DEEPEST POINT: MARIANA TRENCH,
PACIFIC OCEAN
HIGHEST TEMPERATURE: 56.7°C,
DEATH VALLEY, CA
SURFACE AREA: 510,064,472 KM

After a while, the rocket approached something round and shiny. He gasped when he realised it was the moon that he was passing by!

NAME: THE MOON
LIFE SUSTAINABLE: POSSIBLY
AGE: 4.425 BILLION YEARS
HIGHEST TEMPERATURE: 127°C
LOWEST TEMPERATURE: -153°C
THE MOON SMELLS LIKE: GUNPOWDER
IS THERE WATER ON THE MOON: YES,
CLOSE TO THE POLES.

He could see so many
different planets floating
around him and the
bright shiny sun from a
far-off distance.

NAME: THE MOON
LIFE SUSTAINABLE: POSSIBLY
AGE: 4 YES BILLION YEARS
HIGHEST TEMPERATURE: 127°C
LOWEST TEMPERATURE: -153°C
THE MOON SMELLS LIKE GUNPOWDER
IS THERE WATER ON THE MOON: YES
CLOSE TO THE POLES

He took off his seatbelt
and he was able to float
inside the rocket.

What an awesome feeling
Jeremy felt to be floating
inside and his drink
spilling all over.

POWER
FUEL
OXYGEN
AUTO
PILOT
ON
JOY-O-
METER

The rocket then took a turn and headed back to earth.

This was the most happiest moment in his life.

Jeremy knew one day he would be an ASTRONAUT!

Kid's Activity Page – Colour In

About the Author

Sanu is a teacher and resides in Australia. She has always wanted to write a children's book that relates to Science and help evoke interest in young kids on current topics but in a more interesting , imaginative and colourful way.

About the Illustrator

Duane is a cartoonist and illustrator and resides in Drouin, Australia. He has a Bachelor of Fine Arts in Printmaking, Computer Design and Communication and enjoys drawing and bringing images to life.

This book is about a young boy named
Jeremy who takes a trip on a rocket.
The book hopes to intrigue kids about
the outer world and space.

Jeremy is a young boy who loves to play outside. One day he takes a trip on a rocket and experiences an incredible journey.

An exciting story of a young boy's adventure to space.